Raniele Barboza de Souza Pontes
Elisabeth Regina Alves Cavalcanti Silva

Google Earth images for analyzing the dynamics of vegetation cover

Raniele Barboza de Souza Pontes
Elisabeth Regina Alves Cavalcanti Silva

Google Earth images for analyzing the dynamics of vegetation cover

Case study on the fragmentation of forest remnants in a caatinga area

ScienciaScripts

Imprint

Any brand names and product names mentioned in this book are subject to trademark, brand or patent protection and are trademarks or registered trademarks of their respective holders. The use of brand names, product names, common names, trade names, product descriptions etc. even without a particular marking in this work is in no way to be construed to mean that such names may be regarded as unrestricted in respect of trademark and brand protection legislation and could thus be used by anyone.

Cover image: www.ingimage.com

This book is a translation from the original published under ISBN 978-613-9-65377-5.

Publisher:
Sciencia Scripts
is a trademark of
Dodo Books Indian Ocean Ltd. and OmniScriptum S.R.L publishing group

120 High Road, East Finchley, London, N2 9ED, United Kingdom
Str. Armeneasca 28/1, office 1, Chisinau MD-2012, Republic of Moldova, Europe
Printed at: see last page
ISBN: 978-620-7-77404-3

SUMMARY

ACKNOWLEDGMENTS

I thank God, because without him I wouldn't have been able to complete this long journey.

I would like to thank my wife for her support and confidence in me.

I would like to thank my parents for their support and strength in my undergraduate studies, because without them I would not have completed the Agronomy course, and consequently I would not have completed a postgraduate degree.

I would like to thank my professors and colleagues for having contributed with relevant knowledge and for the valuable bond of friendship created.

I would like to thank my advisor for her commitment, patience and help in completing my final work.

"If *where is* important to your business, then geoprocessing is your working tool."

Camara and Davis

SUMMARY

The package of tools currently offered by Google Earth, using high-resolution orbital images, enables detailed observation of practically the entire surface of the planet. The use of these tools offers great potential for observing and researching geomorphology, particularly in terms of recording landscapes and their changes over time, thus providing new perspectives for analysis and complementing the performance of traditional methods already in use. With this in mind, this exploratory study analyzes the dynamics of vegetation cover in an area of Caatinga in the Northeast of Brazil over the course of a decade, using only the visual interpretation of satellite images made available by Google Earth, with the aim of contributing to the use of the service by institutions and ordinary citizens in observing natural and anthropogenic changes on the Earth's surface. It was concluded that between May 2002 and September 2013, in the 16,501 hectares studied, there was a reduction of 495 hectares of vegetation cover, as well as an increase in the fragmentation of forest remnants.

Keywords: Google Earth. High Resolution Images. Visual interpretation. Dynamics. Vegetation cover.

1 INTRODUCTION

In Brazil, the expansion of cultivated areas and land use represent pressure factors on the environment. In this way, the sustainable use of natural resources has become an important issue, due to the increased pressure from anthropogenic activities, leading to an urgent need for continuous monitoring of land use and occupation, with a view to adopting control measures and implementing policies aimed at the sustainable use of the environment, due to the attention given to land use and its economic, social and environmental implications (BECEGATO et al., 2007).

With the intensification of anthropogenic activities, techniques have been developed that have resulted in the establishment of the most diverse forms of land use and occupation (PINTON E CUNHA, 2008). In this context, the authors consider that an integrated analysis of natural attributes, as well as the dynamics of land use in a certain area, contribute to understanding the modification of the landscape and the development of the physical processes related to it. For Ross (2001), integrated studies of a given territory presuppose an understanding of the dynamics of how the natural environment functions with or without human intervention.

Tenedório (1989) states that information on the use of a given space is essential for territorial planning, as it is fundamental to the process of understanding the dynamics and organization of space. In this direction, mapping the landscape units identified from the perspective of their fragility in the face of material conditions and possible human intervention is of valuable importance (ROSS, 2001. In: CRUZ Et. Al. 2009).

The characterization of land use and occupation in time series, when well spatialized and evaluated, is an efficient way of analyzing changes, being determined by periods that indicate strong changes or where we can find a significant degree of evolution of a specific activity or characteristic of the environment (SANTOS, 2003). Land use is defined as the relationship between natural space and people's actions in their environment. Cheng (2003) defines land use broadly as the level of spatial accumulation of activities such as production, transaction, administration and residence with strong dynamic relationships between them. It thus has an economic perspective, in which use is related to the economic and functional activities for which the land is intended.

In this way, the soil, which was once completely natural, undergoes changes due to anthropogenic action and, consequently, the various activities aimed at producing one or more products or services (FAO, 2003).

In this sense, the use of geotechnologies has made important contributions to the monitoring of vegetation cover and land use in recent decades. Among the main tools related to geotechnologies, one of the techniques that can potentially be applied is geoprocessing. Novo (1989) cites some of the applications that this tool can offer in environmental assessment work, such as the assessment of water resources, land use, urban planning, etc. Remote sensing data can be used to extend specific information to a wider spatial context.

Rosa and Brito (1996) point out that the methods used to produce maps and geographical analysis are time-consuming and costly. It is therefore important to use new technologies such as information systems to plan a basin. Satellite images provide a view of the complex and dynamic nature of large areas of the earth's surface.

The integration of GIS with remote sensing techniques has been used in land use planning, where spatial data is integrated with land use data (MENEZES et al. 2009). Thus, using these techniques it is possible to produce maps of land use and occupation in a given area, in time series, in order to assess its evolution (SILVA et al. 2006; BOLFE et al. 2008). Land use and occupation data stored in GIS makes it possible to plan and carry out technical actions for management planning to support policies aimed at its orderly occupation (RODRIGUES, et al. 2011), as well as environmental monitoring of natural resources.

Environmental monitoring is understood as the systematic knowledge and monitoring of the situation of environmental resources in the physical and biological environments, with a view to recovering, improving or maintaining environmental quality (PNMA II, 2009 - 2014). *In other words, it* is a process of data collection, study and continuous monitoring of environmental variables, with the aim of identifying and evaluating - qualitatively and quantitatively - the conditions of natural resources at a given time, as well as trends over time (EMBRAPA, Technological Information Agency).

The development and application of tools that can help analyze the dynamics and monitoring of land use and cover have been the subject of countless studies and research, especially geotechnologies. In monitoring the dynamics of land use, it is essential to use high-capacity systems for processing and analyzing multi-thematic information, such as remote sensing and geographic information system (GIS) techniques (FILHO, 1995). GIS plays an important role in this study because it facilitates the management of spatial information and allows diagnoses and prognoses to be made, supporting decision-making. Remote sensing, on the other hand, has become one of the most effective ways of monitoring the environment at local and global scales due to the speed and periodicity with which it can obtain primary data on the earth's surface (SILVA and MARTINS, 2005).

When it comes to monitoring forest resources in a country as continental as Brazil, orbital images currently stand out as one of the most efficient resources for observing vast landscapes. Advances in technology now make it possible to monitor damage and changes to land cover in almost real time, since the continuous modernization of the geotechnological sector, specifically artificial earth imaging satellites, offers more and more images with better temporal, spatial and spectral resolutions.

In this context, traditional remote sensing techniques have undergone extensive advances to integrate with virtual systems, such as NASA's Google Earth and World Wind platforms, the Google Maps map server, Microsoft's Virtual Earth and Yahoo's map server. These include Google Earth, which is free software whose function is to present a three-dimensional model of the globe, built from a mosaic of satellite and aerial images obtained from various sources (EMBRAPA, 2012).

And given the need to exploit this important tool for environmental monitoring of natural resources, which is Google Earth, this work aims to analyze the efficiency of monitoring the dynamics of soil vegetation cover through purely visual observation and interpretation of high-resolution orbital images provided free of charge by the Google Earth software service.

2 OBJECTIVES

2.1 General objective

To analyze the efficiency of monitoring the dynamics of vegetation cover through the visual interpretation of high-resolution orbital images provided free of charge by the Google Earth software service.

2.2 Specific Objectives

Encourage ordinary citizens and small institutes to use free programs such as Google Earth to monitor morphological changes in the environment, together with other programs such as Quantum GIS and OpenOffice Calc.

To prove that it is possible to monitor environmental aggressions, regardless of the size of the area, especially illegal deforestation, without the need to invest large sums of money in acquiring high-resolution images and GIS programs, or to have extensive knowledge in the area of geoprocessing, which is necessary to work with conventional methods of remote sensing studies.

To carry out a satisfactory analysis using as few tools as possible, and thus contribute to the knowledge that is needed to improve environmental monitoring and inspection systems in Brazil, which can be carried out by any citizen or institution, regardless of their purchasing power or social function.

3 THEORETICAL FRAMEWORK

3.1 Geographic Information Systems (GIS)

Geographic Information Systems are a set of computer programs and procedures that allow the analysis, spatial integration, management and representation of geographic space and the phenomena that occur in it, organized in a spatial database that allows and facilitates the analysis, management or representation of space and the phenomena that occur in it (JONES, 1997; ROCHA, 2007; BLASCHKE; KUX, 2009).

A GIS therefore consists of four elements: hardware, software, information and human resources. In this regard, authors such as OZEMOY et. Al. (1981). stated that GIS represented a set of automated functions that provide professionals with advanced capabilities for storing, accessing, manipulating and visualizing georeferenced information. For Cowen (1988), GIS was a decision support system involving the integration of georeferenced information in a problem-solving environment. For Aronoff (1989), GIS represented the set of procedures, manual or automated, used to store and manipulate georeferenced information.

3.2 Geoprocessing

Geoprocessing can be understood as a set of technologies that aim to collect and process spatial information for a specific purpose. Each geoprocessing application is executed by a specific system, which is called a Geographic Information System (GIS) (SILVA, 2006).

According to Lazzarotto (2003, p.1):

> "Geoprocessing is the set of at least four categories of techniques related to the processing of spatial information: - Techniques for collecting spatial information (Cartography, Remote Sensing, GPS, Conventional Topography, Photogrammetry, alphanumeric data collection); Techniques for storing spatial information (Databases - Object Oriented, Relational, Hierarchical, etc.); - Techniques for processing and analyzing spatial information, such as Data Modeling, Geostatistics, Logical Arithmetic, Topographic Functions, Networks, etc.).); - Techniques for processing and analyzing spatial information, such as Data Modeling,

In this context, the term Geoprocessing denotes the discipline of knowledge that uses mathematical and computational techniques to process geographic information and which is increasingly influencing the fields of Cartography, Natural Resource Analysis, Transportation, Communications, Energy and Urban and Regional Planning. GISs make it possible to carry out complex analyses by integrating data from different sources and creating georeferenced databases. They also make it possible to automate the production of cartographic documents (CÂMARA; DAVIS, 2001, p. 2).

3.3 Remote Sensing Imaging

MENESES, et. al. (2012, p.11, 12) define remote sensing as a technique for obtaining images of objects on the Earth's surface without any physical contact between the sensor and the object. This definition of remote sensing is explicit in stating that the object being imaged is recorded by the sensor through measurements of electromagnetic radiation, such as sunlight reflected from the surface of any object. The authors state that no other type of sensor that obtains images other than by detecting electromagnetic radiation should be classified as remote sensing. Thus, all real-world objects that are at a temperature above absolute zero (-273.15° C or zero Kelvin) can be imaged by remote sensors and the higher the temperature of the source, the greater its radiating power.

The distribution of the intensity of electromagnetic radiation in relation to its wavelength or frequency constitutes the electromagnetic spectrum. Knowing that electromagnetic radiation of each wavelength interacts in different ways and with different intensities with terrestrial objects, one of the most important parameters for defining the characteristics of a sensor are the wavelengths of the images it will acquire. Images are not defined by a specific wavelength, but by small intervals, called spectral bands (MENESES, et. al. 2012, p. 19).

Over the last few decades, there has been a remarkable evolution in the technologies used to observe and measure the Earth. Remote sensing is one of these technologies which, because it receives large investments, is developing at an intense pace. Currently, images from the most diverse types of sensors are available, opening up a huge range of options for work that needs to take into account the geographic characteristics of the places where they occur (TOCHETO, 2010, p. 240).

With this in mind (MENESES, et. al. 2012, p.02) mention the perfect combination of two technologies (artificial satellites and imaging sensors) which, according to the authors, may have been one of the greatest benefits to date of technological development in the service of surveying the earth's natural resources. The authors also mention that what justifies the advance of this technology is its ability to image the entire surface of the planet in a short space of time and in a systematic way, given that a satellite is continuously orbiting the Earth.

These products, presented in specific areas or in a more regional context, allow for efficient diagnoses, propose low-cost solutions and create intelligent alternatives to the challenges faced in the face of the accelerated changes we are seeing in our territory. Remote sensing data has proved extremely useful for studies and surveys of natural resources (INPE).

3.4 Satellite Images in Environmental Monitoring

In recent decades, satellite images have proved to be of great importance in environmental monitoring, especially forest monitoring. Mascarenhas, et al. (2008), analyzing 21 scenes from the Sino-Brazilian Land Resources Satellite (CBERS II - CCD), obtained between July and August 2006, and acquired free of charge from INPE, evaluated images from remote sensing to study the remaining vegetation cover in the Araguaia River Basin, more specifically the Upper and Middle Araguaia, in a total studied area of 120,333.58 km. Using NDVI, they concluded that 74,046.99 km (61.54% of the total area studied) had already been deforested and of the 14,250.1 km of APPs along the rivers, 6,352.56 km, which corresponds to 44.58%, had already been devastated.

GIONGO et. al. (2013), using Spring software, analyzed land cover using images from IRS P6 or

RESOURCESAT-1, with LISS III sensor, orbit 326, point 090, 23.5 m spatial resolution, obtained free of charge from INPE's spatial database and data from a topographic map. The aim of this study was to map land use and occupation and check the suitability of APP areas in the Coqueiros stream sub-basin in the city of Santa Helena de Goiàs-GO using satellite images. The research classified the land uses and coverages at the time of imaging, where agricultural land was the most representative with 73.97%, followed by pasture with 11.63%, forest area with 5.90%, urban area with 3.39%, water surface with 0.70% and other unclassified areas with 4.38%. The survey found good coverage of the APP areas along the stream, with approximately 107.29 of the 128.26 ha required. It was also found that the basin was well preserved, although some points were still in disagreement with the legislation of the time.

3.5 Acquisition of satellite images

In Brazil, it is possible to acquire satellite images free of charge, for example, by accessing INPE's Remote Sensing Data Center, which has the most complete database of orbital images in the country. The service makes available free of charge to registered users the collection of images from the LANDSAT series of satellites since 1973 and the CBERS series since 2003 (TOCHETO, 2010, p. 256). However, these images have the disadvantage of low spatial resolution. LANDSAT, for example, offers spatial resolution of 15 to 120 m depending on the sensor and spectral band used. CBERS, on the other hand, has a spatial resolution of between 20 and 260 m, also depending on the sensor and spectral band (UFRS).

When it comes to high-resolution satellite images, there are currently a number of companies that sell images, which are generally purchased by medium-sized or large companies or by sectors linked to the government. Normally, high-resolution images, especially spatial resolution, come at a high cost. TORLAY and OSHIRO (2010), in a study using Google Earth images, state that high-resolution images are still very expensive on the satellite image market. The authors cite the example of a 192 km scene from the Eros satellite, which costs between R$ 2,000.00 and R$ 3,000.00, or from the Ikonos satellite, which costs between R$ 40.00 and R$ 50.00 per km, or from the Geoeye satellite, which ranges from R$ 50 to R$ 195 per km, with a minimum order of 100 km in the case of Geoeye.

In this light, AYACH et. al. (2012) state:

> "The major current limitation to the widespread use of high spatial resolution images is mainly due to their high cost. However, following the availability of high spatial resolution images on Google Earth and the Stitch Map software, a new trend has emerged for thematic mapping."

Satellite images are not always acquired for the purpose of environmental monitoring, as these images provide other observations and research that are useful in other areas of science, such as geology, agriculture, urban planning, etc. For example, MELGAÇO and FILHO (2003) used the ETM+ sensor on board the LANDSAT-7 satellite to detect deposits of the metallic minerals zinc and lead with great positional accuracy in areas between the municipalities of Itacambira and Monte Azul, in the north and northeast of Minas Gerais. In an agricultural application, MENDES et. al. (2009), also using LANSAT images, mapped and estimated the area planted to soybeans in a region encompassing part of the cities of Tabapora and Sinop in the state of Mato Grosso. Satellite images have proved to be essential for mapping soybeans or any other crop, say the authors. With regard to urban planning applications, BATISTA and BORTOLUZZI (2007), in a study aimed at mapping thirteen cities in the Greater Florianópolis region, state that the use of high-resolution images is appropriate for urban planning applications, since it is possible to identify streets, houses, rivers and other elements of interest to municipal managers.

The images marketed always have a minimum area and swath width at the time of acquisition. With regard to the width of the imaged band, UFRS states that, for example, the SPOT 5 satellite, with a spatial resolution of 5 meters, offers a minimum imaged band of 60 x 60 km; the QUICKBIRD 2 satellite operated by DigitalGlobe, which has a spatial resolution of between 61 and 72 cm, offers a scene width of at least 16.5 x 16.5 km. This means that other landscapes are always imaged that are not of interest to those who are purchasing them, thus allowing the companies selling the images to offer the same product, i.e. the same image, to several interested parties, which can make the cost lower for future buyers. However, the acquisition of high-resolution images is still practically unfeasible for ordinary citizens or small institutions such as schools, small town halls, associations, NGOs, etc.

3.6 Visual Interpretation of High Resolution Satellite Images Applied to Land Cover Studies.

The techniques of visual interpretation of satellite images or photointerpretation are part of the data analysis system in remote sensing (NOVO, 1995, p. 6. in PANIZZA and FONSECA, 2011). Thus, not every study involving a remote sensing product requires the analysis of electromagnetic reflectance, pixel by pixel, although this is the most widely used method. In many cases, it is possible to make observations in order to extract the necessary information visually.

Visual interpretation can be defined as the act of examining images in order to identify objects and judge their significance. As such, the interpreter's task is not only to identify and delineate objects precisely, but to try to define regions that present uniformity in terms of composition and appearance. The interpreter generalizes to define spatial units that make up the subject of interpretation (LUCHIARI, A. 2010 in VALE, T. S. 2014). Also according to PANIZZA and FONSECA (2011), visual interpretation is procedural and goes through stages that are essential for understanding the phenomenon. Stages such as identification, determination and interpretation are fundamental and decisive for the interpreter to be able to extract the maximum meaning from the images.

During the interpretation process, the following activities are carried out almost simultaneously: detection, recognition, analysis, deduction, classification, accuracy assessment. In addition, visual interpretation is based on seven image characteristics in the information extraction process, such as: tone/color, texture, pattern, location, shape, shadow and size (NOVO 2008 in AYACH, et. al 2012).

In visual interpretation or photointerpretation, the first stage is a simple reading of the image. At this point, the user almost intuitively makes a correlation between the object observed and another known object. Next, the user develops mental processes (deductive or inductive), even if the image only reveals a partial view of the object, the top of a tree or the roof of a house. Finally, the user creates correlations between the elements determined in the image and draws up interpretative hypotheses. For example, a pool of water observed on flat ground may indicate the existence of impermeable soil (BARIOU, 1978, p. 43 in VALE, T. S. 2014).

In this perspective, Leite et. al. (2014) in a study that analyzed changes in land use in the city of Montes Claros, between 2000 and 2011, using high spatial resolution satellite images from the Ikonos-II satellite, with a spatial resolution of 1 meter in panchromatic, and images from the WorldView-2 satellite, which has a spatial resolution of 0.50 meters in panchromatic. Using ARC. GIS 10.0 software to vectorize the features, they understood the dynamics of land use and the rate of growth, as well as the location of this transformation. The research concluded that in the period analyzed, the occupied area of the city expanded by 27.2%, and that the main driver of this growth was the residential class, concentrated mainly on the southern and eastern outskirts of the city. The conclusions of the study provide a general understanding of the changes in the urban system and can be used as a subsidy by public managers when defining management policies.

Valério Filho and Belisàrio (2012), in a study on the adequacy of urban zoning legislation in the municipality of Sao José dos Campos-SP, used high-resolution orbital images to analyze the dynamics of urban occupation in the region comprising the Pararangaba stream sub-basin in 1997, 2003 and 2008. The aim of the work was to analyze and interpret high-resolution images in order to map the temporal evolution of the urbanization process and assess the level of soil sealing in the sub-basin region. Topographic maps on a scale of 1/10,000 were used, as well as thematic maps obtained from the town hall containing digital cartographic and thematic material in SPRING (.spr) format for the municipality's planimetric and altimetric data, color aerial photographs on a scale of approximately 1/10,000 and images from the Quick Bird sensor, which has a panchromatic spatial resolution of 0.60 m. These images were interpreted directly on the computer screen. The research concluded that there was a decrease in drainage capacity along the watershed, increasing the likelihood of flooding events downstream of the watershed, and also concluded that there was inconsistency in the zoning law regarding the classification of urbanized areas.

3.7 About the Google Earth Program

Formerly known as Earth Viewer, Google Earth was developed by the company Keyhole, which was acquired by Google in 2004 (renamed Google Earth in 2005).

Since then, significant changes have been implemented on this platform, especially with regard to improving the quality of resolution, georeferencing accuracy and updating images. These improvements have led to concern about national security issues in some countries. However, many GIS experts are still skeptical about the applicability of the Google Earth platform for scientific purposes. At the beginning of 2015, free access to the Pro version of Google Earth was released, with additional features compared to the basic version (EMBRAPA, 2012).

The kml language (Keyhole Markup Language) is the interface between Google Earth and the user. It is an open standard markup language that allows geographic data to be displayed in geonavigators, such as points, lines, polygons and images. In its compressed form it is called kmz (EMBRAPA, 2012).

OLIVEIRA and SAITO (2012) draw attention to the unreliability of the images and vector layers posted by users on Google Earth, whose authenticity and veracity cannot be audited. When it comes to Google Earth, the authors state: "*there is* no criterion to certify the accuracy of positioning or any other mechanism to certify the veracity of authorship or any other attribute of the information posted".

Despite the multiplicity of sources, dates of acquisition and resolution of the images that make up the Google Earth platform, several studies attest to the viability of using it to monitor the Earth's surface.

3.8 The potential of Google Earth images

SOARES et al. (2010) analyzed the quality of Google Earth images for the municipality of Pato Branco, PR, and concluded that they met the Cartographic Accuracy Standard (PEC) compatible with the 1/30000 scale.

SILVA and NAZARENO (2009) analyzed the accuracy of the imagery made available by Google Earth for Goiânia, GO, and found that it complied with PEC class A at a scale of 1/5000 and had a reliability level of 90%. In similar analyses, Oliveira et al. (2009) demonstrated that the accuracy of the images of Sao Leopoldo, RS, is compatible with the 1/15000 scale.

In a study aimed at mapping land use and land cover, AYACH et. al. (2012) used high-resolution images from the GeoEye satellite of the region of the Córrego Indaià Hydrographic Basin in the state of Mato Grosso do Sul, available on Google Earth. The interpretation key for the study was developed using 3 (three) interpretive elements: Tone/Color, Texture and Shape. Regarding the results of their work, the authors state:

> "The use of high-resolution images from the GeoEye satellite extracted from Google Earth in the preparation of the Land Use and Vegetation Cover mapping of the Córrego Indaia- MS Hydrographic Basin was satisfactory, where it was possible to interpret information that allowed 5 (Five) thematic classes to be mapped. This result demonstrates the potential of Google Earth images as a thematic base for mapping land use and land cover."

Torlay and Oshiro, (2010) georeferenced high spatial resolution images available on Google Earth of the Barao Geraldo district, in the municipality of Campinas-SP, in order to generate products for land use and occupation applications in the region, including visualizations, localization, spatial data processing and area calculation. The study concludes that the classification of use and occupation generated was more accurate than that generated from LANSAT 5 images, which, although free, have low spatial resolution. The authors also mentioned that the advantage of Google Earth is that it provides high-resolution images free of charge. However, the authors warned that the correct image processing procedures should be carried out using GIS tools in order to obtain the best results.

4 WORKING METHODOLOGY

4.1 Characteristics of the Study Area

The first stage of this work was to choose the area to be studied, which necessarily had to be covered by high spatial resolution satellite images in the Google Earth program, in natural colors, with a time interval between scenes of more than a decade and without any cloud cover, since all observation for data interpretation had to be purely visual.

An area located in the Sertao region of the state of Pernambuco was chosen for the study, specifically in part of the municipalities of Ouricuri and Bodocó. After choosing this area, using Google Earth's polygon vectorization tool, we made a virtual demarcation of the polygon that was the object of the study.

The area studied (figure 01) has a rectangular shape with an area of 16,051 Ha (sixteen thousand and fifty-one hectares) and a perimeter of 53,119 m (fifty-three thousand and one hundred and nineteen meters). It is georeferenced and calculated in the UTM Projection Plane, Datum SIRGAS 2000, Zone 24 South.

Figure 01. Situation Map and Details of the Area Studied

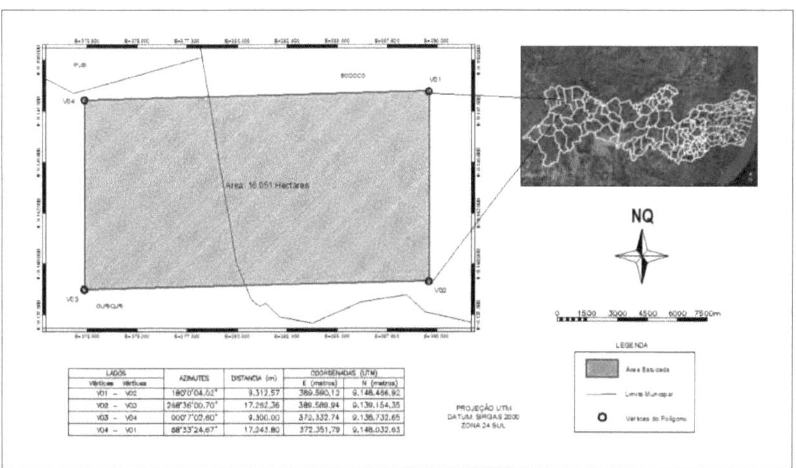

The area was selected because it had the necessary characteristics for the work to be carried out in the Google Earth program, i.e. high spatial resolution, natural colors, a temporal coverage interval of more than a decade and no cloud cover. Parts of three images were used in the research (Figure 02 and Figure 03). One scene was taken on October 5, 2002 from the DigitalGlobe collection and the other two are from 2013, one on September 3, 2013 and the other on August 28, 2013, both from the Astrium collection. There is therefore a time gap of just over eleven years between the observations.

Figure 02. Scenes used in the research from 2013

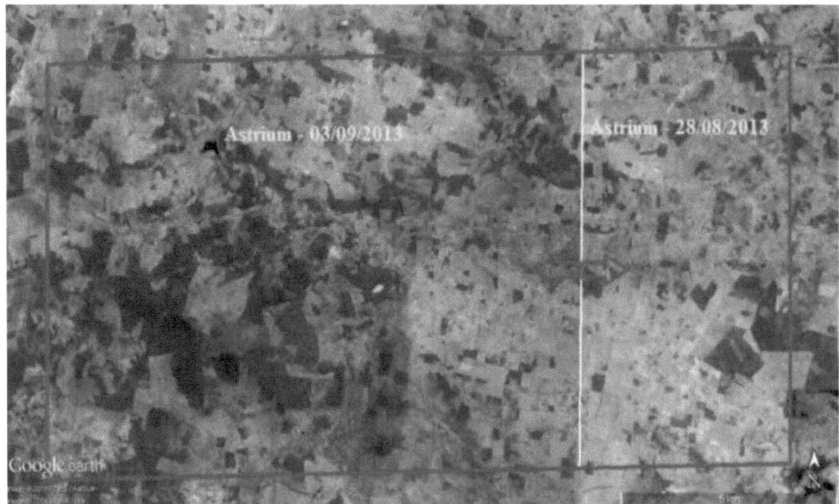

Source: Google Earth

Figure 03. Scene used in the research from 2002

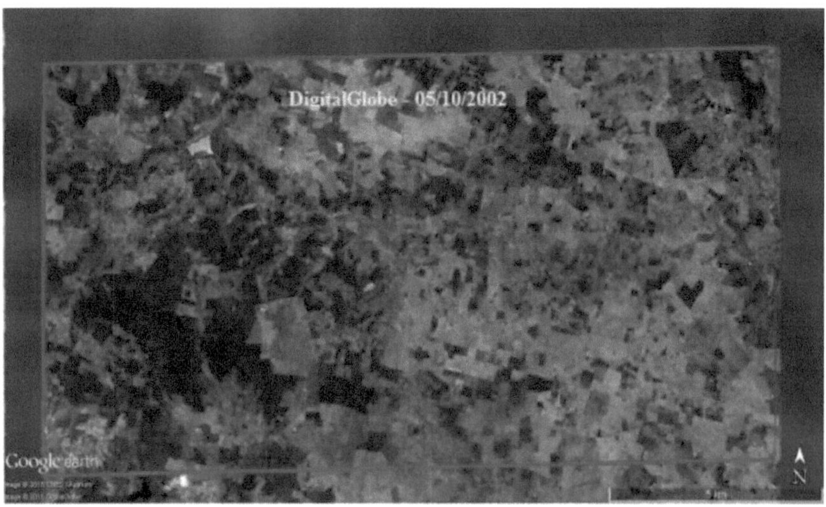

Source: Google Earth

4.2 Determining Control Points

Next, using the "historical images" tool in Google Earth, a detailed observation was made in order to choose objects to be used as control points. The aim of determining control points was to analyze the georeference accuracy of the satellite images and to estimate the maximum scale of work that could be used in studies with the scenes used in the research. These control points had to be fixed objects, easily identifiable in the field, well distributed throughout the research area and present in all the images used in the work.

After careful observation on the Google Earth screen, seven objects were selected which were white in color and circular in shape and which were previously interpreted as cisterns, since all these points were located next to a masonry building and persisted in the same place over the time span of the different scenes, in this case just over eleven years. Figure 04 shows the spatial distribution of the selected control points in relation to the study area. Figures 05 and 06 show how one of the control points used appears in the different satellite images and figure 07 shows a current photograph of the control point represented in figures 05 and 06.

Figure 04 - Spatial distribution of control points

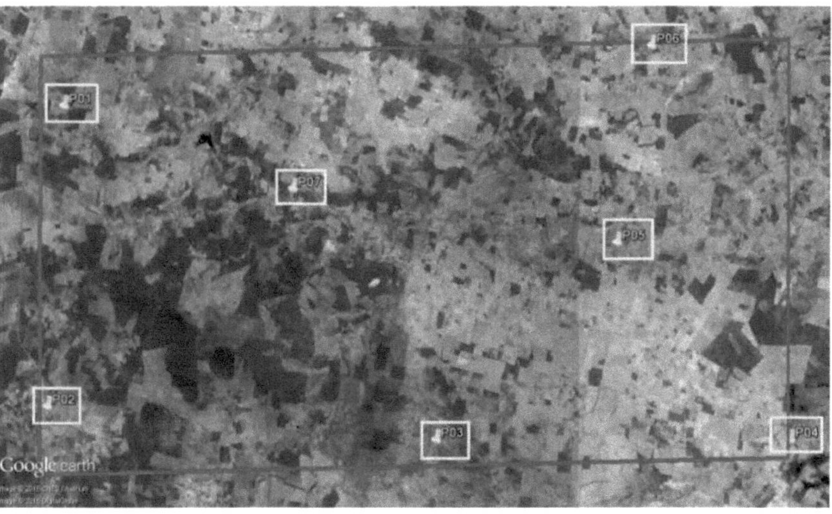

Source: Google Earth

Figure 05. Control point P02 from figure 04 in the 2013 scene.

Figure 06. Control point P02 from figure 04 in the 2002 scene

Figure 07. Photograph - cistern used as control point P02

Source: Author

The coordinates of these objects were then collected directly on the computer screen. All the points were then exported to a Garmim Etrex Legend HCx GPS navigation device to be located in the field. As previously interpreted, all the points were located and confirmed to be cisterns.

Since the aim of this work was not to georeference the images precisely, but only to carry out an environmental study of the dynamics of vegetation cover, in order to estimate the scale of the work, high-precision tracking with geodetic GPS devices was not used at the control points, but only tracking with navigational precision.

Therefore, after locating each cistern used as a control point in the field, they were occupied and the respective coordinates were collected by positioning the device in the center of the object, using the same Garmim Etrex Legend HCx GPS navigation device.

4.3 Vectorization of vegetation cover

Once the coordinates of the control points had been collected in the field and the polygon delimiting the research area had been defined, using the same Google Earth polygon vectorization tool, the areas covered with vegetation that were located within the researched polygon were vectorized. The vectorization procedure of the areas covered by vegetation was carried out on both the 2002 and 2013 scenes. The vectorizations generated layers that were saved separately in KML format.

In order to be vectorized as a polygon, the vegetation fragment analyzed in the scene had to be completely separated from other vegetation patches. In addition to the dividing line caused by areas that had already been cleared and used for farming, the main roads seen in the satellite images were considered to separate these patches. Figures 08 and 09 show the two conditions that generate forest fragmentation in the region.

Figure 08. Fragments generated by deforestation for crops.

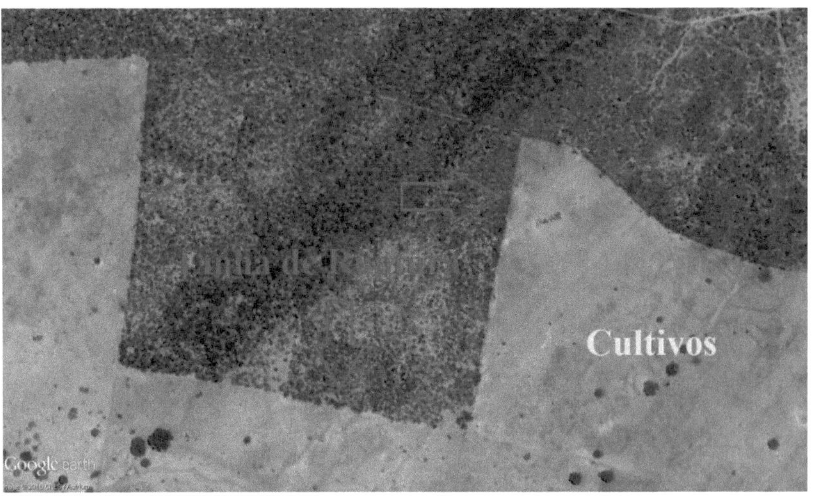

Source: Google Earth

Figure 09. Fragments generated by roads.

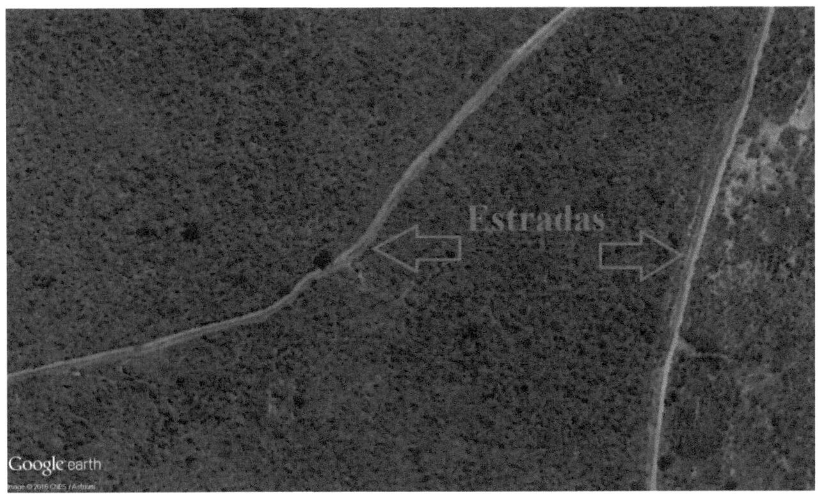

Source: Google Earth

4.4 Vectorized Data Analysis

Once the areas had been identified, vectorized and the files saved, they were then imported into Quantum GIS, version 1.7.4, and new files were generated, now in Shapefile format for all the vectorizations. In Quantum GIS, using the "Area" tool in the Field Calculator of the Attribute Table, the area values in square meters in the UTM Plane of each polygon were calculated and the resulting files were generated and saved again in Shapefile format. Saving files in Shapefile format generates four other file formats in addition to a file with an "shp" extension. One of the files has the extension "dbf" which is the Database file, where all the calculations and data contained in the Attribute Table are saved. The "dbf" extension has the characteristic of being able to have its files opened in one of the programs in the free OpenOffice package, OpenOffice Calc, which corresponds to Excel in Microsoft's Office package.

Since one of the aims of the work was to simplify the methodology of the resources used, practically all the spatial analysis and other calculations were done in OpenOffice Calc, with only the calculation of the area of the vectorized polygons being done in Quantum GIS. Generally, all these analyses are carried out in geoprocessing programs that have specific tools for these functions, but they present the obstacle of being complicated to use and not very intuitive. However, because OpenOffice Calc's platform is based on the Excel platform, most of its tools and functions are practically the same. This helps make it possible for more people to use the program for these purposes, since Excel is part of practically all basic computer courses and is present on most desktop computers in Brazil, both at home and in the workplace. dbf' files can also be opened in Excel and perform the same calculations and analyses, but they were not used in this work because they are not free software, even though Excel is better known and used than *OpenOffice Calc.*

The respective calculations and analyses described below were carried out in *OpenOffice Calc*: quantification and summation of the overall total of the areas of the fragments covered by vegetation; quantification and percentage of the area of the fragments by area intervals, where nine different intervals were determined to be analyzed, which are: (0.1 to 1.0; 1.0 to 10; 10 to 25; 25 to 50; 50 to 100; 100 to 200; 200 to 300; 300 to 400 and 400 to 500 hectares); comparison of land cover dynamics in the time interval studied between the 2002 and 2013 scenes and generation of graphs and tables of the results obtained.

5 RESULTS AND DISCUSSION

The aim of this study was to evaluate and quantify the percentage of vegetation cover and its dynamics over a period of just over a decade, based solely on visual interpretation of images provided free of charge by the Google Earth program. These images are panchromatic, with high spatial resolution and an RGB composition with natural colors. For this reason, they have a large scale of detail and significant roughness, which made it possible to visually distinguish between exposed soil, areas of agricultural crops and vegetation cover. In other words, even without an analysis of the electromagnetic reflectance per pixel using conventional processes, it was possible to analyze and quantify the characteristics of the land cover of the entire area studied.

A simple observation of the images of the site shows that the total area covered by vegetation is significantly less than the total area covered by exposed soil. Based on this observation, it was concluded to vectorize only the features identified as vegetation cover, to the detriment of exposed soil cover, thus facilitating the identification of the features. However, by calculating the difference between the area of the polygon studied and the sum of the areas of their respective areas covered by vegetation, the area of exposed soil was determined. In this study, vegetation patches smaller than 1000 m (0.1 hectares) were not considered to be vegetation fragments, i.e. all polygons representing vegetation cover had an area in the UTM plane greater than 0.1 hectares.

Looking at the vectorized polygons in the 2002 scene (Figure 10), which represent the fragments of areas covered by vegetation, it is difficult to see the magnitude of the dynamics of these fragments when compared to the 2013 scenes (Figure 11). This is due to the size of the area studied, which is relatively large for an efficient naked eye analysis of these changes. To solve this problem, Quantum GIS and OpenOffice Calc were used to perform the necessary calculations. The combined use of these programs proved to be very efficient in analyzing the dynamics of these fragments in vector data generated in the Google Earth program.

Figure 10. Map of the 2002 scene with vectorized areas

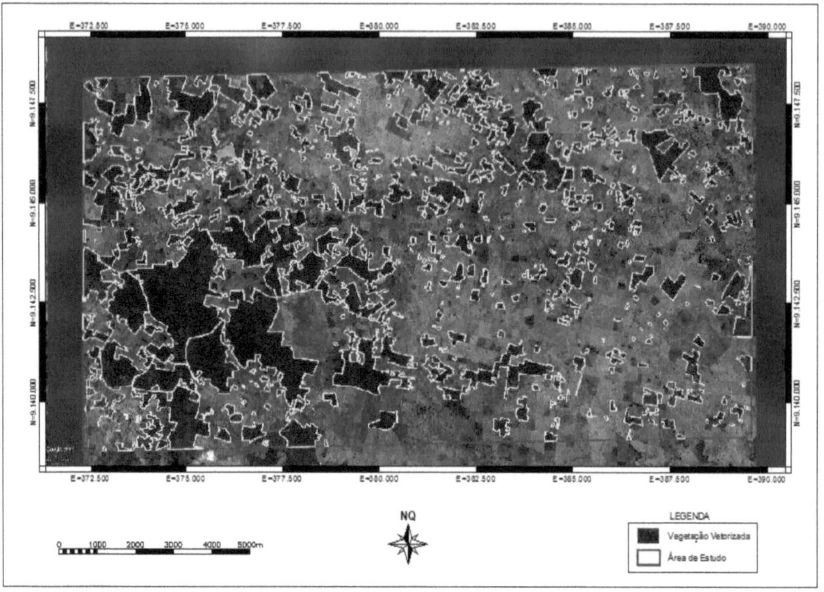

Source: Author

Figure 11. Map of 2013 scenes with vectorized areas

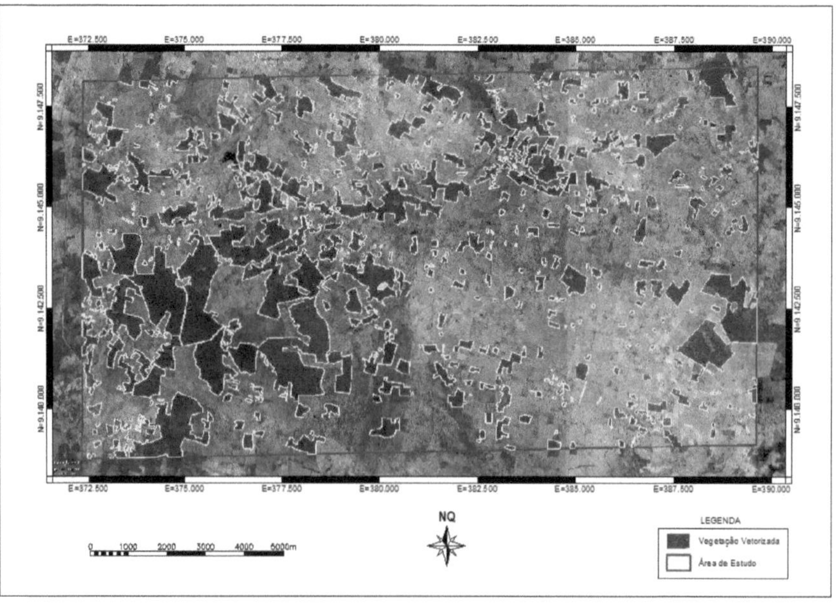

The Google Earth program does not provide information on the spatial resolution of its images, but by observing the details, it is visually estimated that they are images with a spatial resolution of less than two and a half meters, more specifically around 1 meter, and are therefore considered to be of high resolution. For example, the scenes show cisterns, which are objects with a diameter of less than 2.50 meters.

Because the area is located in the region of the drought polygon in the Semi-Arid Zone of Northeast Brazil and because it has Caatinga as its native vegetation, on the dates of the image generation the vegetation was completely dry and the areas of agricultural crops had completely exposed soils, with no or little vegetation, since the scenes are from the driest period of the year, which is between the months of June and January. This made it easier to differentiate between the features analyzed.

As the Google Earth program uses images in their raw format, i.e. without having previously undergone orthorectification, it is extremely important for the use of these images to know the level of geometric distortion they present, since in applications such as cartography, mosaics, geographic information systems and others, the detection and monitoring of spatial changes in land features requires data with good spatial accuracy.

The images used in the research have good geometric accuracy and are perfectly suitable for other correlated studies that do not require high precision. Table 01 shows the comparison between the coordinates of the control points collected in the field and the coordinates of the images obtained virtually on the Google Earth screen.

Table 01. Coordinates of control points and geometric difference.

UTM coordinates, SIRGAS 2000, Zoua 24 South										
Coutiole Point	Image 2002		Image 2013		Coordinates Collected with GPS Navigation		Difference in distance between Image 2002 and o Navigation Point		Difference in distance between Image 2013 and Navigation Point	
	This	Norie	This	Norie	This	North	This	North	This	North
P1	372.828.78	9.146.754.57	372.830.95	9.146.754,77	372.839,49	9.146.750.95	10,71	3.62	8,54	3.82
P2	372.455.01	9.140.064.55	372.455.21	9.140.065.55	372.459.86	9.140.061.48	4.85	3.07	4.65	4.07
P3	381.411.87	9.139.349.10	381.412.54	9.139.350.33	381.415,77	9.139.347.04	3.9	2,06	3,23	3,29
P4	389.585.13	9.139.374.00	389.589.21	9.139.375,52	389.586,61	9.139.371,97	1,48	2.03	2.6	3,55
P5	385.582.28	9.143.751.71	385.586.15	9.143.748.74	385.584.21	9.143.748.91	1.93	2.80	1.94	0.17
P6	386.351.64	9.148.123.18	386.357.91	9.148.119.88	386.356.66	9.148.120.21	5.02	2.97	1.25	0.33
P7	378.135.62	9.144.918.09	378.137.24	9.144.918,89	378.139,51	9.144.917.45	3,89	0.64	2,27	1.44

Using the data on the differences in distances between the coordinates of the control points collected in the field and the coordinates observed in the satellite images, it was possible to estimate the scale of the scenes studied. It can be seen that the biggest difference is in the EAST coordinate of point P1 with 10.71 meters for the 2002 scene and 8.54 meters also in the EAST coordinate of point P1 in the 2013 scene. The rest of the differences are all practically less than 5 meters. Adding these differences to the error resulting from the accuracy of the navigation GPS, which in theory is less than 10 meters, but in practice can be well below this value, we conclude that the maximum ideal scale is 1/10,000, i.e. there is no positional error significantly greater than 20 meters.

In this study, the areas covered by vegetation identified in the satellite images as crops were not considered to be forest fragments. This can be seen in Figures 12 and 13, where the plants are distributed in rows and parallel lines. Given the region's farming practices, it is likely that the majority of these are fodder palm plantations used to feed cattle.

Figure 12. Satellite image of the site with crops in parallel lines.

Figure 13. Photograph of Palmal in the Municipality of Bodocó.

After all the analysis and calculations, the study found, using the methods used, that the 16,051 hectares studied had a total vegetation cover in the 2002 scene of 4,646 hectares and an exposed soil cover of 11,405 hectares, which represents 28.15% and 71.85% respectively. In the 2013

scenes, vegetation cover was 4,151 hectares and exposed soil cover 11,900 hectares, corresponding to 25.86% and 74.14% respectively (figure 14).

Figure 14 - Land cover graph

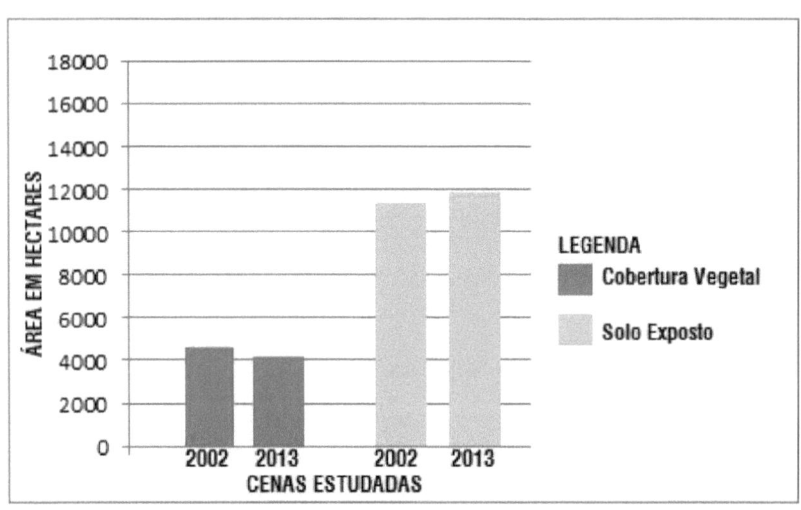

Source: Author

It can be seen, then, that in just over 11 years, there has been a decrease in the area covered by vegetation of around 2.29 %, which corresponds to 495 hectares of deforested areas, and it is very likely that most of this was illegal deforestation, since in the region studied there is no culture of deforestation being carried out legally, coupled with little monitoring by the responsible bodies with regard to attacks on native vegetation in the municipalities of the region studied.

When analyzing the dynamics of vegetation fragments, the research showed that there had been an increase in their number. The 2002 scene showed a total of 493 patches of vegetation. The 2013 scene showed a total of 519 patches, which corresponds to an increase of 5.27% compared to 2002. The increase in the number of patches, which was caused by deforestation in the region, points to a greater level of degradation of natural resources. The process of environmental fragmentation exists naturally, but has been intensified by human action. This action has resulted in a large number of environmental problems (RAMBALDI and OLIVEIRA, MMA, 2003).

This should be seen by society and government officials as a worrying and highly relevant fact. Changes in the structure of the fragments were also observed over the period, i.e. a decrease in the number of large fragments and an increase in the number of smaller ones (Table 02). Conservation biology knows that the larger the area of continuous forest formation, the more balanced the biotic relationships. The opposite occurs in fragmented areas due to the impacts of the process known as Edge Effect, which increases as the number of fragments increases.

Table 02 - Vegetation cover and dynamics in the fragments.

YEAR 2002

IAF (ha)	QF	%QF	SAF (ha)	PCAE
400 - 500	1	0,2	431,972	2,69
300 - 400	2	0,41	608,428	3,79
200 - 300	0	0	0	0
100-200	5	1,01	643,063	4,01
50- 100	14	2,84	996,859	6,21
25-50	12	2,43	436,941	2,72
10-25	35	7,1	572,675	3,57
01 - 10	243	49,29	866,518	5,4
0,1 - 01	181	36,71	89,9411	0,56

YEAR 20013

IAF (ha)	QF	⁰ZoQF	SAF (ha)	PCAE
400 - 500	0	0	0	0
300 - 400	0	0	0	0
200 - 300	2	0,39	513,048	3,2
100 - 200	6	1,16	881,795	5,49
50 - 100	9	1,73	595,564	3,71
25 - 50	16	3,08	568,957	3,54
10-25	46	8,86	709,56	4,42
01-10	235	45,28	778,959	4,85
0,1-01	205	39,5	103,373	0,64

IAF: Fragment Area Range

QF: Number of Fragments

%QF: Percentage of Fragments

SAF: Sum of Area of Fragments

PCAE: Percentage of Coverage in Relation to the Area Studied

Source: Author

In the 2002 scene, 2.69% of the area studied was covered by patches of vegetation between 400 and 500 hectares and 3.79% by patches between 300 and 400 hectares. This fact was not repeated in the 2013 scene, which did not show any patches with areas in these ranges.

Another relevant fact was the number of fragments between 0.1 and 1.0 hectare, which are considered to be very small patches and harbor less biological diversity when compared to large fragments. These have increased over the period studied. The 2002 scene had 181 fragments, which together totaled 89.94 hectares, corresponding to 0.56 % of the total area studied. The scene from 2013 had 205 fragments, which together totaled 103.37 hectares, corresponding to 0.64% of the total area studied.

6   FINAL CONSIDERATIONS

The decrease in the area covered by vegetation, together with the decrease in the number of large fragments and the increase in the number of small fragments, may have had a significant impact on local biodiversity.

The conversion of areas covered by vegetation in the region into deforested areas may have occurred mainly for the use of new agricultural crops, since dairy farming is practiced in the region studied, especially in the municipality of Bodocó, which is one of the largest milk producers in the state of Pernambuco.

The possibility of visually observing and interpreting high-resolution satellite images available on Google Earth provides important support for monitoring the dynamics of vegetation cover. By simply navigating the program's screen and manipulating the historical images tool, it is possible to detect areas that have undergone morphological changes. In this way, any person or institution will be able to monitor the changes that have taken place on the earth's surface, especially attacks on forests. This will enable the detection of various environmental crimes and the consequent possibility of reporting them to the competent authorities.

Because the Google Earth platform is made up of mosaics of georeferenced images, even though they are raw images, i.e. uncorrected, it makes it easier to find where the changes are likely to occur.

The continuity of the service offered by Google in terms of the free availability of high spatial resolution satellite images in the Google Earth application will in future provide vast areas covered with scenes from different imaging dates. This could become a powerful tool for recording and monitoring the dynamics of landscapes, mainly due to the ease of use, popularity of the application and free access. In addition to this tool, Google has another service known as Global Forest Watch (GFW). This is an interactive online forest monitoring and alert system that provides near real-time alerts showing suspected locations of recent tree cover loss. GFW is free and allows anyone to create personalized maps, analyze forest trends, subscribe to alerts or download data for their local area or worldwide. However, the service needs studies to prove its efficiency and application in Brazil.

Observation of the images used, combined with the few other tools in the other programs also used in this research, made it possible to quantify and analyze the changes that occurred in an area of 16,501 hectares, which can be considered a significantly large area. However, in the case of smaller areas, it is possible to carry out similar operations in an even simpler way, since Google Earth offers not only vectorization tools for features such as lines and polygons, but also area calculation tools. In other words, you can carry out observations and studies using just Google Earth.

REFERENCES

BECEGATO, V. A.; FEREIRA, F. J. F.; CABRAL, J. B. P.; FIGUEIREDE, O. A. R.; NETO, S. L. R.. **Monitoring land use and occupation in the area of influence of the municipality of Fazenda Rio Grande - Metropolitan Region of Curitiba - PR**. RAE GA - O Espaço Geogràfico em Anâlise, Curitiba, n. 14, p. 217-227, 2007. UFPR Publishing House.

BLASCHKE, T.; KUX, H. **Advanced remote sensing and GIS**. 2. ed. Sao Paulo: Oficina de Textos, 2009. 303 p.

BOLFE, E. L.; MATIAS, L. F.; FERREIRA, M. C. **Geographic Information Systems: A contextualized approach in history**. Geografia (Rio Claro). Rio Claro, v. 33, n. 01, p. 69-88, 2008.

CHENG, J. **Modelling spatial & temporal urban growth**. 2003. 203 f. Theses (Doctoral in Geographical Sciences) - Utrecht University. Utrecht, 2003.

FAO, **Food and Agriculture Organization of the United Nations**. Planning for sustainable use of land resources. FAO land and water bulletin 2. Rome: FAO, 2003.

MENEZES, M. D.; CURI, N.; MARQUES, J.J.; MELLO, C.R.; ARAÙJO, A.R..

Pedological survey and geographic information system in the evaluation of land use in a hydrographic sub-basin in Minas Gerais. Ciência e Agrotecnologia. Lavras, v. 33, n. 6, p. 1544 - 1553, 2009.

NOVO, E. M. **Remote sensing: principles and applications**. Sao Paulo. Edgard Blucher, 1989. 309p.

PINTON, L. G.; CUNHA, C. M. L. **Evaluation of the dynamics of linear erosion processes and their relationship with land use evolution**. Sao Paulo, UNESP, Geosciences, v.27, n. 3, p. 329-343, 2008.

ROSA, R.; BRITO, J. L. S. **Introduction to geoprocessing: geographic information systems.**

Uberlândia: Editora da Universidade Federal de Uberlândia. Uberlândia. 1996. 104p.

ROSS, J. L. S. Geomorphology: environment and planning. 6th edition, SP, Contexto, 2001.

SANTOS, M. A. **Construction of scenarios in a GIS environment to evaluate land use changes induced by hydroelectric plants in the agricultural region of Andradina**. 2003. 153 f. Dissertation (Master's Degree in Agricultural Engineering) - State University of Campinas, Faculty of Agricultural Engineering. Campinas. 2003.

TENEDÓRIO, J. A. **Design of land use and evolution charts by interpreting vertical aerial photography. Almada**: methodological example. National Institute for Scientific Research. Center for Geography and Regional Planning Studies: FCSH-UNL, 1989.

CRUZ, L. M.; PINESE JÚNIOR, J. F.; SILVA, T. I. **Analysis of the environmental fragility of the Glória stream basin in Uberlândia-MG using cartography and geoprocessing techniques**. Available at:

http://www.geomorfologia.ufv.br/simposio/simposio/trabalhos/trabalhos_completos/eixo1/05 2.pdf. Accessed on 02/02/2016.

LAZZAROTTO, D. R. **What are geotechnologies**. 2002.

SILVA, M. S. **Geographic information systems: elements for the development of distributed geographic digital libraries**. 2006. 167 f. Dissertation (master's degree) - Universidade Estadual Paulista, Faculdade de Filosofia e Ciências. Marilia. 2006.

SILVA, E. R. A. C.; MELO, J. G.; GALVINCIO, J. D. **Identification of Areas Susceptible to Desertification Processes in the Middle Section of the Ipojuca Basin - PE through the Mapping of Vegetation Hydric Stress and the Estimation of the Aridity Index**. Revista Brasileira de Geografia Fisica, v. 4, n. 3, p. 629-649. 2011.

LEITE, M. E.; PEREIRA, A. M.; NOBRE, B. A.; MARTIN, A. S. **Monitoring the dynamics of urban land use in Montes Claros/MG using high spatial resolution images**. Caminhos de

Geografia Uberlândia, v. 15, n. 51, p. 172-180. 2014.

RODRIGUES, T. R. I.; PERES FILHO, A.; ARAÙJO, G. K. D. **Monitoring the dynamics of land use and occupation in the lower course of the Sao José dos Dourados River, SP - Area of influence of the Ilha Solteira and Très Irmaos hydroelectric reservoirs**. In: Simpósio Brasileiro de Sensoriamento Remoto - SBSR, XV, 2011, Curitiba, Anais, INPE, 2011, p.6932.

SILVA, R. C. C.; MARTINS, A. K. E. **The use of CBERS-2 images to quantify and qualify degraded areas in the Rio Formoso Project in the municipality of Formoso do Araguaia-TO.** In: Simpósio Brasileiro de Sensoriamento Remoto - SBSR, XII, 2005, Goiânia, Anais, INPE, 2005, p. 1083-1089.

ROCHA, C. H. B. **Geoprocessing: transdisciplinary technology**. Juiz de Fora. UFSJ, 2007. 220 p. 20.

EMBRAPA Technological Information Agency. **Environmental Monitoring**. Available at: http://www.agencia.cnptia.embrapa.br/gestor/cana-de-acucar/arvore/CONTAG01_73_711200516719.html. Accessed on 18/01/2017.

GUIMARAES, D. P.; PIMENTA, F. M,; LANDAU, E. C. **Google Earth-SIG Map Server Integration and Environmental Monitoring.** EMBRAPA. Sete Lagoas. ISSN 16791150. December, 2012.

OLIVEIRA, C. H.; SAITO, C. H. **The image of the landscape and the landscape of the image: the system for acquiring, processing, hosting and integrating information on environmental resources (SAPHIRA)**. Espaço & Geografia, Brasilia, v. 15, n. 2, p. 385-405, 2012.

OLIVEIRA, M. Z.; VERONEZ, M. R.; TURANI, M.; REINHARDT, A. O. **Google Earth images for environmental planning purposes: an accuracy analysis for the municipality of Sao Leopoldo/RS**. In: Simpòsio Brasileiro de Sensoriamento Remoto - SBSR, XIV, 2009, Natal, Anais, INPE, 2009, p. 1835-1842.

NAZARENO, N.R. X and SILVA, L. A. **Analysis of the cartographic accuracy standard of**

Google Earth images, using the image of the city of Goiânia as the study area**. In: Simpósio Brasileiro de Sensoriamento Remoto - SBSR, XIV, 2009, Natal, Anais, INPE, 2009, p. 1723-1730.

SOARES, M. C.; RUARO, T. A.; AGUIAR, C. R. de. **Quality Control of the Map Base of the City of Pato Branco in Google Earth Software**. Synergismus Scyentifica UTFPR, Pato Branco, v. 5, p. 28-30. 2010.

OZEMOY, V. M., SMITH, D. R. AND SICHERMAN, A. **Evaluating computerized geographic information systems using decision analysis**. Interfaces, San Francisco, v. 11, n.5, p. 92-100, 1981.

COWEN D.J. **SIG versus CAD versus DBMS: what are the differences?** Available at: https://docs.ufpr.br/~firk/personal/Carto_Digital/SIG%20CAD%20DBMS.pdf. Accessed on: 03/02/2017.

SILVA, E. R. A. C.; MELO, J. G. S.; GALViNCIO, J. G. **Identification of areas susceptible to desertification processes in the middle section of the Ipojuca basin - PE through mapping of vegetation water stress and estimation of the aridity index**. Revista Brasileira de Geografia Fisica, v. 4, n. 3. p. 629-649. 2011.

ARONOFF S. (1989) Geographic Information Systems: A Management Perspective. Ottawa: WDL Publications.

MASCARENHAS, L. M. A.; FERREIRA, M. E.; FERREIRA, L. G.; **Remote sensing as a tool for environmental control and protection: analysis of the remaining vegetation cover in the Araguaia River basin**. Sociedade & Natureza, Uberlândia, v. 21, n. 1. p. 5-18. 2009.

CBERS - China-Brazil Earth Resources Satellite. Available at: http://www.dgi.inpe.br/siteDgi/portugues/satelites.php. Accessed on 17/01/2017.

VALERIO FILHO and BELIZARIO P. R. **High resolution orbital images applied to the study of the adequacy of urban zoning legislation in the Pararangaba stream sub-basin**. Geography Teaching & Research. Sao José dos Campos, v. 16, n.1. p. 173-177. 2012.

GIONGO, P. R.; AZEVEDO, L. C.; COSTA, R. A.; SILVA, P. C.; MARCOMINI, A. M. **Mapping land cover and conflict with Permanent Preservation Areas through remote sensing**. Enciclopédia Biosfera, Centro Cientifico Conhecer - Goiânia, v.9, n.16, p.1409, 2013.

Global Forest Watch (GFW). Available at: http://www.globalforestwatch.org. Accessed on: 20/01/2017.

CÂMARA, G.; DAVIS, C. In: CÂMARA, G.; DAVIS, C.; MONTEIRO, A. M. V. **Introduçâo à Ciência da Geoinformaçâo**. Sao José dos Campos: INPE, 2001. p. 2.

TORLAY, R.; OSHIRO, O. T. **Obtaining images from Google Earth for classifying land use and occupation**. 2010. Available at: https://www.embrapa.br/busca-de- publicacoes/-/publicação/872526/obtencao-de-imagem-do-google-earth-para-classificacao-de- uso-e-ocupacao-do-solo. Accessed on: 21/01/2017.

UFRGS. **Sensors and Orbital Platforms**. Available at: http://www.ufrgs.br/engcart/PDASR/sensores.html. Accessed on: 23/01/2017.

MENESES, P.R.; ALMEIDA, T.; ROSA, A. N. C. S.; SANO, E. E.; SOUZA, E. B.; MESQUITA; FILHO, J.; BAPTISTA, G. M. M.; BRITES, R. S. - **Introduction to remote sensing image processing**, 2012, p. 11, 12, 19.

SAUSEN, T. M. **Remote Sensing and its applications to natural resources**.

Available at: http://www.inpe.br/unidades/cep/atividadescep/educasere/apostila.htm#tania. Accessed on 23/01/2017.

FANIZZA, A. C.; FONSECA, F. P. **Techniques for the visual interpretation of images**. GEOUSP - Espaço e Tempo, Sao Paulo, n. 30, p. 30 - 43, 2011.

VALE, T. S. **Google Earth as a methodological procedure in the pedagogical practice of Geography in Elementary School II**. 2014. 171 f. Dissertation (Master's in Geography) - Pontifical Catholic University of Sao Paulo, Sao Paulo, 2014.

OLIVEIRA, C. H.; SAITO, C. H. **The image of the landscape and the landscape of the image: the system for acquiring, processing, hosting and integrating information on environmental resources (SAPHIRA)**. Espaço & Geografia, Brasilia, v. 15, n. 2, p. 385-405, 2012.

MENDES, F. S.; MORAES, E. C.; ARAI, E.; DUARTE, V.; MOURA, Y. M. **Soybean monitoring using satellite images: Tabapora and Sinop regions, Mato Grosso**. In: Simpósio Brasileiro de Sensoriamento Remoto - SBSR, XIV, 2009, Natal, Anais, INPE, 2009, p.287-291.

RAMBALDI, D. M.; OLIVEIRA, D. A. S. **Fragmentation of ecosystems: causes, effects on biodiversity and recommendations for public policies**, Brasilia, MMA/SBF, 2003.

PNMA II. **Component: Institutional Development - Sub-component:**

Environmental Monitoring. National Environmental Program II - PNMA II, Phase 2, Brasilia, MMA, 2009.

I want morebooks!

Buy your books fast and straightforward online - at one of world's fastest growing online book stores! Environmentally sound due to Print-on-Demand technologies.

Buy your books online at
www.morebooks.shop

Kaufen Sie Ihre Bücher schnell und unkompliziert online – auf einer der am schnellsten wachsenden Buchhandelsplattformen weltweit! Dank Print-On-Demand umwelt- und ressourcenschonend produziert.

Bücher schneller online kaufen
www.morebooks.shop

info@omniscriptum.com
www.omniscriptum.com